Thinking Outside the Boombox

ISBN: 978-1-7334560-0-5

Cover art by Marcus Kwame
www.marcuskwame.com
Instagram: marcus_kwame

Thinking Outside the Boombox

By

Justin "The JUSE" Ellis

Table of Contents

Chapter 1:

Master of Your Own Ceremony

Page 1

Chapter 2:

...Like Deciphering Graffiti

Page 57

Chapter 3:

B-Boys and B-Girls, Be Men and Be Women

Page 99

Chapter 4:

The World is Like a Mixtape

Page 121

Chapter 5:

God Isn't a DJ, So Stop Making Requests

Page 177

Chapter 6:

Word to Your Mother Earth

Page 221

Chapter 1

Master of Your Own Ceremony

1

You can always make more money, but you can never make more time. Don't let people steal your time. Protect it. Be stingy with it. It's not replenishable.

Notes

2

You have different versions of yourself: the pauper, the worker, the soldier, the king, and the god. They all deal with your concept of self, characteristics, and self-perception: how you move, how you interact with people, how you affect the world around you, etc. Think about these different stages and their place in the world, the respect they receive, the impact they have on a world or society, the decisions they make. Elevate yourself to your god state.

Notes

3

The reason conscious people don't connect with most people is because “most people” don't have anything to connect to. Most people are just empty shells.

Notes

4

There's knowledge, the fact that you comprehend a concept or idea or are aware. Then there's understanding, which is recognizing when and how to use the knowledge you have. Lastly, there's wisdom, actually applying your understanding in real life. We should all strive for wisdom.

Notes

5

If you could, would you choose to change how people perceive you or how you perceive yourself?

Notes

6

A wise man does not argue with fools.

Notes

7

Everyone knows the world is fucked up, so you will inevitably have to do some fucked-up shit to survive, but don't become a fucked-up person.

Notes

8

People are often indifferent or expect something negative to happen, almost like it's inevitable or it's an expectation. Think about the moments in your life where things started to go down and spiral and how you forced yourself to accept it on some level. Why not have that same, or even greater, obsession that things work out in your favor; that you expect good things to happen? Why can't a negative thing really just be your misinterpretation of a positive thing? Think of how much you could accomplish if "negative" things had no effect on you and you didn't sulk.

Notes

9

A man with ideas, a plan, and ambition is very dangerous. Be dangerous.

Notes

10

Are the most successful people the ones that can ask for help? Are you someone who wants to figure it out or someone who asks for help so you can move on? Do you continue to try and try and ignore your current limitation or do you try, but are willing to seek support?
Are the most successful people also those who are not afraid of failure or rejection?

Notes

11

Time alone is time well spent.

Notes

12

In a kingdom of fools, the clown is king.

There is no pride in being the least ignorant person in the room. Instead, place yourself around people who challenge you to be greater, not people who are ruled by stupidity nor people for whom average is acceptable.

Notes

13

To obtain your goals, have a clear vision of the ending, then your mind will naturally reverse-engineer how to accomplish it. It's difficult to start at zero and work your way linearly to a place you've never been. The mind is great at filling in blanks.

Notes

14

Perception is deception of your own mind.

Notes

15

Sarcasm just seems like a way to not be genuine. If you're emotionally mature, you don't have to mask how you feel with sarcasm, because you can be open and honest.

Notes

16

A step in any direction is progression. Stagnation is what kills progress.

Notes

17

Don't live in absolutes or extremes. Instead, find your balance in everything.

Notes

18

Love is an art. War is a sport.

Notes

19

You can't rationalize with someone who is irrational.

Notes

20

Any emotion that is worth having is worth expressing. So, if someone upsets you, tell them. Don't run around mad at someone, but not tell them how you feel. You won't resolve anything by keeping it in.

Notes

21

Never make decisions when you're hungry or horny.

Notes

22

You might be abusing alcohol if: You drink alcohol because you're thirsty You drink warm alcohol You keep alcohol in your car You drink alcohol alone

Notes

23

To be a king or queen, you must first be a servant.

Notes

24

As with Christianity, and other religions, we're told to believe we're flawed and imperfect, but that's a construct; it's a leash to limit our thinking and our potential. It's like the Iceberg Slim story of how he tricked his bottom bitch into staying longer by making her think she made a mistake when Iceberg orchestrated the whole thing. So, if we stop believing the leash, what is our true potential?

Notes

25

You are no longer a slave when you're on a path of healing and resolution. As long as someone can push a button (i.e. call you a "nigger") and you react how they expect you to, then you're still a slave and have a mentality of someone who is mentally crippled. No one has domain over my actions or emotions, or my universe, but me, so I am a slave to nothing.

Notes

26

There are two types of problems people have in life: those they can control and those that they have no control over. Neither is worth stressing over. If you *do* have control, then make a change. If you *don't* have control, then there's nothing you can do about it, so don't let it stress you.

Notes

27

Who decides what the right way to do anything is? We're all just beasts who all have different wants and ideas, but the same needs. Take back your choice to choose your own path.

Notes

Chapter 2

...Like Deciphering Graffiti

28

If someone has something that you desire, they have power over you.

Notes

29

The truth doesn't care about your feelings or opinion.

Notes

30

The logic for not consuming meat or animal products is flawed if you're doing it to protect living things. Plants are alive while they're in the ground or on a tree, but when you uproot them or pluck them from the tree that supplies their nourishment, they're no longer able to grow or reproduce or anything; therefore, you're killing plants.

Where is the threshold? What living things are ok to kill? There are plants that have been proven to have memory or can feel sensation, communicate between each other, or even eat living creatures, or grow towards light, or even underground fungi networks. So, what is the criteria for being alive?

Notes

31

An idea is a contextualized thought.

Notes

32

Why is time divided up into 24-hour segments? Who decided that? Why not 100 or 4? You can divide it up in an infinite number of segments/time periods. If there were less time periods, people would be happier and more active and productive; they would spend less time watching the time and more time doing.

Notes

33

Since there are animals that have abilities humans don't, like sonar, night vision, and so on, it's reasonable to assume that some animals may have abilities that humans can't even observe because of the limits in our senses.

Notes

34

Is someone having passion just someone having a prolonged or consistent emotion towards something, whereas someone else may only have a fleeting feeling?

Notes

35

Creativity often comes from adversity, so the lack of creativity in Hip-Hop could be interpreted as a sign of progress, meaning that the new generation is not as stressed as previous generations (...or it could also be that everyone is medicated, so everyone is numb to the adversity).

Notes

36

If a mirror is made of glass, and glass is made of sand, and so is man, then we are all essentially mirrors and reflections of whatever created us.

Notes

37

It is said that life is mathematics -- we can understand it in math terms, but it is also chemistry. Everything is the result of chemical reactions; elements interacting with other elements. Even thought, ideas, dreams are made up of an element; we're just not able to quantify or substantiate them (yet), but they are a specific combination of elements, like a rock, a cloud, or water or a star, or anything else ...we are chemists and can create once we understand that.

Notes

38

Negative energy kills. Positive energy heals.

Notes

39

We're all made of energy, and energy cannot be created nor destroyed, only change in form. So, when we die, our energy just changes in form.

Notes

40

The same way an atom has protons, neutrons, and electrons, human emotion is either positive, neutral, or negative, and furthermore, people's spirits have a baseline of acting in one of those three states. So, are we like atoms, acting and reacting according to the charges and other atoms around us?

Notes

41

We're confined by the limitations of our own brain. Our brain is electrical signals transmitting, but how do we know that, because it is our brain telling us what our brain is doing. The brain can't actually define itself, and no one can know the universe or existence outside of what our brain tells us. We can't think beyond what we can't think. Every thought or sensation is only chemicals and electricity in the brain.

Notes

42

Words with no substance are like a boat with no oar. They will follow the wave and the current, without intention or direction of their own.

Notes

43

Can humans really take credit for anything they're able to do? Even with things we work hard to get good or better at, it's really because we have the wiring to enjoy working at it and/or the propensity to want to work at it. Some people have the ability to persevere when attempting to accomplish a goal. For example, some people have the capacity to understand math, some people have the capability to work at being good at it, while some people's minds are not set up that way at all, and just can't do it, when it's really all in the way we're wired; even things we think are our own doing, it's really because our brains were wired in a way that allowed for us to be able to do them.

Notes

44

It's liberating to be labeled as crazy, because when you're crazy, you no longer have limitations or constraints. Crazy is just a way of saying "I don't understand you and it scares me."

Notes

45

One day, emotions, like anger or confusion (or their frequencies) and other things we can't observe, that aren't tangible, will be harnessed and used as a weapon. A thunderstorm could be created, but of greed or rain of fear ...that means love or confidence could also be harnessed.

Notes

46

Where faith has failed me, logic has saved me.

Notes

47

Words are just sounds and symbolic scribbles, so how is it that we allow a series of sounds or organized lines to offend us? Anyone's hate or negativity is only as powerful as we allow it to be. The context we give these, otherwise meaningless, concepts is how we give them their power.

Notes

Chapter 3

B-Boys and B-Girls, Be Men and Be Women

48

Fellas, it seems to me that you only call a woman a "bitch" when you can't control her. Who said she is for you to control?

It really means she has no confidence in your ability to lead her.

Notes

49

There are three parts or legs of a healthy relationship: there's physical connection, attraction and sexual chemistry; then there's intellectual, a mental connection where you're at similar places intellectually and are intrigued and challenged by the other person; finally, there's a spiritual connection, this is where you have a good vibe and energy with the other person. Any disorder in a relationship can be attributed to one of these three facets.

Notes

50

Is it better to love someone the way you want to love them or the way they want to be loved?

Notes

51

It's interesting that adults tend to have a more difficult time making new friends; it was easier to make friends when you were a child. I think it's because children seek commonality and adults pick out characteristics and find excuses to not fuck with someone. Children can say, "Hey, you like video games; I like video games. Let's be friends." Adults say, "She gets her hair done at some place" or "he's a Cowboys fan. I can't be cool with him." It's good to choose positive people to surround yourself with, but if you're looking for friends, try finding common ground on something.

Children also do not prejudge.

Notes

52

There is a fine line between love and an addiction.

Notes

53

For my women, think of conversations in this 2/3rds rule: say a conversation with your man has 99 words to share per hour. Your man is gracious enough to give you 66 of those words and he'll only keep 33 for himself. And if you use up your 66 in the first five minutes you need to wait until he uses his 33 before you talk again or it resets at the next hour. If he doesn’t use his 33 then let him reset the conversation on his time and reengage you.

Notes

54

Ask yourself if the people in your life are adding value to your life. In a healthy relationship there is a mutually agreed-upon symbiosis. Also, ask yourself what are you bringing to the table; what value do you have in the relationship. Are you giving more than you're taking (and it doesn't have to be monetary; it could be time, support, fun, love, etc. and it doesn’t have to be 50/50, just an agreed-upon trade that both parties are comfortable with)? Toxic relationships have an imbalance or misunderstanding of reciprocity.

Notes

55

Don't bother breaking up with someone and then later getting back together. The sequel is rarely as good as the original. Either stay together and work it out or if you feel it can't be worked out, then end it.

Notes

56

As parents, your first responsibility is to protect your child; as they get older, your job is to warn them and guide them, but don't let them fall; and, as they enter adulthood, you offer advice, but you allow them to fall, but you are there to pick them up when they holla.

Notes

57

Challenge yourself and your partner to take time just observing each other. Act as if the other person wasn't around; what would you honestly be doing? Try it. Without talking at all. You can learn so much by watching.

Notes

Chapter 4
The World is Like a Mixtape

58

I find it odd that people attach themselves to sports teams (even schools) and athletes, so much so that they allow themselves to get upset at people's kids and are emotionally invested to the point they will be disrespectful to an opposing team - - people who either get paid to play a game or kids who play for fun. Either way, it's a game. Why don't we cheer for other leaders the same way; why do we care so much more for a game that others are playing, for entertainment, than we do the welfare of society and our community? I see little difference between serious sports fans and cult members. Furthermore, there are whole channels and even networks built around other grown men and women talking about these other people who are just playing a game. It all seems a little absurd to me.

Notes

59

Society is emotional. People respond and react based on feelings, and because it's fleeting, arbitrary, not concrete or tangible, it creates confusion and disarray and is distracting. Trump’s tweets are an example of creating chaos by distracting people with emotionally charged rhetoric.

Notes

60

The consciousness of White America is narcissistic. That's how Trump won office. He represents the inflated ego of White Americans. That's why they have a false sense of superiority. They feel like they're losing America (something that isn't theirs to begin with). There's the gaslighting; there's the lack of morals and empathy. They deflect and project, etc. ...And just like a narcissist, they will never admit fault and will continue to do whatever they can to keep the status quo, the supply, with no intention of actually resolving anything.

So, the solution is not to talk or wait for them to help; it's to accept that they have this warped view of reality, but know in your heart that it's not true and live in truth, but they won't change. Don't invest energy into proving you're right or them wrong. Instead, just move accordingly.

Notes

61

Hip-Hop is where it is because White America (the majority) doesn't understand the culture and just magnifies and extracts the pieces they find interesting, like the 808, the gang life, drug dealers, autotuned, etc. ...the things that excite them or stereotype us or that demean us, but also the parts that they can copy. They can't copy black intelligence or black empowerment, but they can imitate us in a minstrel show.

Notes

62

Human nature in modern history, especially this generation, is afraid of conflict. That's why there are all these gender fluidity laws, etc. The most famous people are those who are not afraid of conflict, Trump, Kanye, and so on, because it's a trait most people envy and admire. It's the reason social media is so popular, because people can hide behind the screen, and why trolling and cyber bullying is so pervasive. This is why reality shows are popular; people tune in for the drama and live vicariously through the cast.

Notes

63

What we're seeing in Hip-Hop culture is that it is incredibly adaptive. While most cultures stay the same and adhere to traditions until they die out, Hip-Hop culture is constantly evolving and changing.

Notes

64

Is a factor in determining what will be popular in music based on what seems to be a perversion and hyper-maturity of where the children are currently at in society (for example: sexualizing teen singers, kids talking about love, kids talking about violence and drug dealing, and so on)? This has been a true trend for every generation.

Notes

65

The whole double standard and gender equality is a result of women wanting to be treated as equal, but I don't think women and men are equal. Genetically, they are not the same, (for instance, why I won't hit a woman, why women get half the money in a divorce, but men typically don’t, etc.). When we say genders are equal, we blur the line. When we blur the line then there leaves room for gender fluidity.

If men and women were equal, there would be half women in the NFL, so, no, I don't think they're equal, and I don't treat them as such. I still open car doors for women; I pay for the first, second, and third dates; I walk on the outside when on a street or sidewalk. Women are special in their own way; they have capabilities men don't, and vice versa; men can't get pregnant.

I believe strongly in fairness, but not equality. For the women that want to be "equal," start asking men out on dates and paying for all the dates, opening doors for men, and opening jars, and I bet before long you’ll wish you had a “man.”

Notes

66

Regardless of your sexual preference or how you choose to identify, there is still precisely one way to create a new life: You still need a healthy sperm and egg cell.

Love who you want freely; be who you want to be freely, but don't pretend that it's "normal."

Notes

67

Soon, social media companies will find a way to monetize likes and follows, so users and businesses and, of course, the social media companies can get paid for them. They would have to find a way to verify or validate the likes and follows so someone can't hack the system and automate likes. The likes would have to come from a legitimate, unique source.

Notes

68

There will be a new crime, and a wave of crimes, of falsifying someone doing a crime by creating a video that makes it look like someone did the crime or with a hologram. With technology, it's easy to superimpose someone's face on another person in a video.

Notes

69

Think of all the things the powers-that-be are able to and could do without the masses noticing because everyone has their head down looking at their phones or on social media. Politically, economically, Project Blue Beam; they could start the NWO and most people wouldn't even notice. It's all a diversion.

Notes

70

A more accurate way to identify people and their place in society and their stance, more so than the voluntary ethnic identification forms, is by culture: Hip-Hop, white supremacist, black supremacist, vegan, nationalist, surfer, socialite, farmer, rural, etc. People can identify more accurately with a culture than with race or ethnicity (for example, a white person who was born and raised in the hood, might identify more with the urban culture).

Notes

71

While I appreciate anyone willing to sacrifice for what they believe, I don't accept that there is honor in killing innocent people or starting or engaging in conflict with a country that is less of a threat to my rights or my life than our own government. You want to earn my respect; let me see you fight for your community. Fight the real oppressors. Fight the mother fuckas stealing tax money out of my checks to pay off the national debt; fight the police who are abusing their power and killing innocent blacks; fight the people who raise the interest rates on credit card and student loan debt so much so that people are financially drowning; attack the banks that deny people loans based on their ethnicity; attack the mother fuckas who gentrify neighborhoods and displace people; attack the bigots and racists, those are the real terrorists.

Notes

72

When it comes to politics, don't endorse a person, endorse certain values. Humans are flawed, but respect, honor, truth, and love have yet to be corrupted.

Notes

73

Everything we need to cure our ailments can be found naturally. It all already exists on the Earth; we just need to figure out what it is, cultivate it, and use it in the right away or determine what combination of ingredients to use. Whether it's heartburn, or cancer, or foot fungus, I believe the Earth was created with everything we need. Our earth consists of the vast majority of the elements that are the building blocks of the universe. So, it has to be that some element or mix of elements can be harvested from things that are found naturally on Earth to heal us.

Notes

74

What does it actually mean when someone says that someone acts "black" or acts "white?" I reject that there is a single way that blacks or whites act, yet the presumption is that if someone acts black, they're being ghetto or ignorant. No one ever says someone is acting black and that person is being intelligent or investing or being responsible. Conversely, if someone is acting or talking "white," then the implication is that the person is articulate or educated or proper. Why do we accept that being black is being foolish?

Notes

75

Because blacks are a minority, we're always fighting *against* someone or something. We're forced to be reactive; it can be overwhelming, but we need to get to the point of fighting *towards* a solution.

Notes

76

If you're a member of the LGBT Community or just "different", stop announcing it. If you want to be treated equally, stop pointing out your differences if it really doesn't matter; just be a good person and promote the aspects of your character you DO want to be known for.

Notes

77

If we're to believe that people who are on death row deserve to die, and that they cannot be rehabilitated, then instead of using tax money to house them indefinitely, why not have a program where the inmates could repay society in a different way?

Say, either allowing them to serve in the military, on the most dangerous missions (like during war), and after a certain term, they could be released into society (if they survive), which is better than having innocent people and children fight. Or we could allow them to volunteer to be human experiments for medications and vaccines and live fairly normal under supervision, since the best way to know how medications affect humans is to test on a human, but be upfront with what they're being tested on, no need to lie, and allow them to choose that option. They could have an opportunity to contribute to society and an opportunity for redemption.

Notes

78

Contrary to popular opinion, the Black Community IS evolving in many ways because we're more open. Before it was taboo or frowned upon to speak out on certain things, now people are becoming bolder and more open about everything: sex, rape, mental health issues, violence, so the community is getting better, because you have to be open to be able to heal, we're going away from the slavery of silence. The Black Community is often at the forefront of technology, embracing social media and galvanizing across the internet.

Notes

79

If someone requires a handicap parking sticker because they have a disability that prevents them from driving normally, it would seem to me that they shouldn't be on the road because they are a danger to themselves and others. Either you can drive safely or you can't.
Furthermore, why in the world would someone who has a disability that prevents them from driving regularly need a sports car?
We would need to provide them alternate transportation.
We would need to expand the public services greatly.

Notes

80

Everyone should be free to pursue what brings them peace and joy, but I don't agree that I or society should have to change or acquiesce to your feelings. Societal standards should be based on fact, not emotion or feelings. Feelings are on a spectrum, facts are not. I shouldn't have to change my vocabulary because of how you feel. If that's how you feel, then you're free to change. Feelings are not measurable or observable, but facts are.

(For instance, the transgender or non-binary bathroom debate is really about forcing society to accept how these people feel, instead of setting clear guidelines of biology)

Notes

81

From the early 2000s to current, music changed formats and was no longer physical (CDs were the last physical representation of music). With music becoming nothing more than a file or an abstract idea and intangible, people appreciated music much less, so talent was replaced with a product.

Notes

82

Racism, prejudice, and every other form of oppression are possible through the majority having control or influence. What if the population was completely equal, where would whites be? Where would blacks be? The oppression all goes back to who is the majority (not exclusively race) and can align like people. Blacks are represented well for only being 13% of the population. What if blacks were actually half of the population? To hold a greater portion of the population, we need to make babies! Support black families and fertility. Fuck Planned Parenthood. We need our own clinics.

Notes

83

It is the responsibility of those who know better to correct the behavior of those who don't. That is how a society progresses - through leadership and setting standards.

Notes

84

There is this false prevalent narrative that there is scarcity. It's not true. Society will have you believe that there is not enough money or resources to go around, but there is. There's an unlimited supply; it's just not allocated or cultivated appropriately. And with things that seem to be in short supply, there's always enough ingenuity and ideas to go around to find alternatives.

Notes

Chapter 5

God Isn't a DJ, So Stop Making Requests

85

Claiming a religion is like saying Burger King is your favorite fast food restaurant, but you've never tasted any of the other restaurants. We know fast food is not healthy, in general, but all the fast food restaurants have something tasty when you have a craving. Arby's fries, Wendy's burger, Burger King's fish sandwich, and so on. Every religion has something attractive about them, but real sustenance, is a home-cooked meal, that comes from actually meditating and building a relationship with the Source.

Notes

86

Because religion and spirituality are such difficult concepts for humans to grasp, the vast majority just don't think about it. They just find something and go with it, so they don't have to think anymore. There is only a small portion of people who actually continue to think deeply and challenge and question the norm until they find something they are satisfied with.

Notes

87

The body is merely a prison for the spirit, and freedom is death.

Notes

88

People spend all this energy and emotion putting their faith into some entity or character or concept that is mere conjecture, claiming it's helped them. But what if they focused that faith within themselves and things that are tangible? Just imagine what humans could achieve. It wasn't that imaginary being supporting you; it was YOU. Just aim and focus your faith.

Notes

89

Philosophy asks the questions religion pretends to answer.

Notes

90

I've only seen "God" do two things reliably: raise the sun every morning and cause human suffering.

Notes

91

If you find a religion or belief system that makes sense to you, then I suggest you go with it. If it helps you sleep at night, and helps you through the day and through hard times, and provides peace, and helps you get closer to your Source, then adhere to those ideologies. However, if you have even one question, then I suggest you search for answers.

Notes

92

Believe in yourself; anything else is pagan.

Notes

93

When it comes down to it, we all are seeking an understanding and relationship with whatever created us. The name doesn't matter: God, the Big Bang, the Source, ...who cares? I doubt any of us are accurate in our nomenclature.

Notes

94

Faith is accepting your fate, that you don't have to be in control, and you're ok with the result.

Notes

95

In the story of the Tree of Knowledge, I think the reason there was forbidden fruit, forbidden knowledge, is because God didn't want humans to get too close to the sun (like the story of Icarus) meaning too close to the truth or the knowledge of God. For those who want the knowledge, they're punished by being tormented mentally with the troubles of the world, wanting to learn more, and feeling like they need to do something about it. Ignorant people are happy because they're dumb, not necessarily in the world, but in truth. Seekers of truth are mentally tortured, while people who don't question what they're told to believe are content.

Notes

96

Who does God believe in?

Notes

97

I have so much faith in my connection to the Source that I don't need to pray. I know that all my needs will be met and I know all of my desires are accounted for, so I have nothing to ask for. My responsibility is to continue to grow and follow the divine path and I will be satisfied.

Notes

98

If you believe God is all powerful, then isn't it just as possible that God can be all of the Gods? So, couldn't God present Himself in different ways that people can accept?

Notes

99

If the objective was for humans to ultimately be as "righteous" as possible, then why would God make humans so flawed and incapable of perfection? An omnipotent God could have created anything and could have created people to be perfect. Why would a God create a flawed creation and then punish the creation for not being perfect? ...unless there's a benefit for that God, more so than for a finite being or creation.

Notes

100

Any type of benevolent figure needs an antagonist. A "good" God needs a devil to be the scapegoat or else people wouldn't have a path to Him and wouldn't have a need for a God Who should be both praised and blamed. There needs to be a dichotomy.

Notes

101

Faith without action is merely wishing.

Notes

102

Why debate someone on their personal relationship with the Source? It's their relationship to have. Religion is divisive because people argue over individual feeling and over something no one can prove nor disprove.

Notes

103

Atheists and religious people, alike, all believe that we come from a Source that is bigger and more powerful than us; whether it's the Big Bang or a god, it's still a source that determined our existence.

Notes

104

The more I read and challenged myself about religion, the more I realized I didn't believe any one religion completely, but I liked and could take out a bit from every religion or philosophy. So, instead of being 100% anything, I'm 3% of this, 20% of that, 10% of this religion. I am an amalgamation of everything I've learned and believe. We'll all the sum of what we believe.

Notes

105

I don't see much difference between an alien and what some people call “God” - a being not of this Earth, who is more intelligent than humans.

Notes

Chapter 6

Word to Your Mother Earth

106

I seriously hope that there is life more intelligent than humans in this vast universe. It would be pretty sad if we, in all of our idiocy, were the pinnacle of intelligence.

Notes

107

What if aliens put the different ethnicities and cultures on different continents every few thousand years, and Antarctica is next and that's why there's a clandestine government interest? Or perhaps, there was already a civilization on Antarctica and it's being kept under wraps. Could this be why artifacts and history overlap, but span different timelines, because cultures were actually living in and around each other and growing together and interacting?

In this case, there would be some isolation because they were planted in different areas, but interacted because they were free to travel and migrate. There are similarities in pyramids and many other things across cultures. Maybe crystal skulls have something to do with it. Maybe aliens were the first to inhabit Earth, then Africans, then Asians, and so on. Maybe even dinosaurs too, but then the Earth had to be reterraformed. Where do "African Americans" come in? African Americans whose ancestors were American slaves don't really look like Africans.

Notes

108

What if the Earth is the result of several terraforming attempts, and this is why there is so much variety of terrains and habitats? That could explain the Ice Age and dinosaur extinction. That could also explain how so many animal and plant species exist. Some survived; some didn't.

Notes

109

What if mythological animals actually existed and had different senses and abilities that we currently can't observe? And maybe those abilities were harnessed by ancient civilizations, like perhaps dragons or unicorns had telekinesis and the Egyptians harnessed it to build the pyramids, but harnessing their abilities hurt the animals and killed them off. That would be no different than whale poaching or hunting elephants for their ivory.

Notes

110

What if every major Earth cataclysmic event marks a new pollination of species by aliens and that's why science continues to discover new species and other microscopic organisms?

Notes

111

A huge percentage of DNA is not used or, at least, we don't understand the use. What if DNA is the code parameters that the Creator had to work within? And what if we become more advanced once we crack the code and are able to understand more of our DNA? What if some stereotypes are actually true, because they were programmed differently into the various groups?

Notes

112

What if the Earth is half of a globe, and there is a separate opposite side, like a yo-yo or zygote splitting, all within a closed environment or dome? If this were the case, both theories of the Earth being spherical and a plane could simultaneously be correct. This could explain how scientists were both right in observation of the Earth, the Coriolis Effect, flight navigation, compasses, gyroscopes, the path and pattern of the sun and how it's going around a sphere, but if we follow the path, it's a straight line, and so on. Things on the ground we would observe as a straight line and things in the sky may appear to have a circular arch.

Maybe the other side is a different dimension, or where Atlantis is, or aliens, or dinosaurs or where the Third Reich was trying to go through Antarctica to (where Admiral Bird spoke of).

Notes

113

What if everything is happening all at once; all of our spirit's or energy's life cycles are happening all at once? The past, the present, and the future are all happening at the same time, but we are only able to observe and experience in instances and memories and premonitions.

Notes

114

Maybe like a zygote, the universe is expanding and multiplying, and the Earth and our universe are just inside one cell, and the other universes are just other distant cells.

Notes

115

Everything comes from darkness and blackness. Color can't exist without black; the absence of anything can't create something (even white people can't create color genetically). Matter, color, or whatever, is all a part of a whole or sum. With this, even the Big Bang exclaims that everything was born of a totality and elements spread. The Bible even says, "God said 'let there be light'" ...*LET* there be light on the first day, yet the sun wasn't created until day 4.

Notes

116

God ultimately represents what we came from and how we came to exist. God represents where our ideas originate, where our spirits originate, where our characteristics originate. For religious people it's an omnipotent being or beings; for atheists, it's the Big Bang from nothing ...but the fact that one of them is correct is equally absurd. For either, there is something that has always existed or there was absolute nothingness and all that we know (and don't know) came from that. Both assertions are mind-boggling.

Notes

117

Maybe we haven't made contact with extraterrestrials because we're not intelligent enough. Maybe what we think of as advanced (like using a hammer, fire, manipulation of electricity, nuclear energy, or whatever) is all primitive in the scope of the universe. So, maybe we're just too "dumb" to communicate with the thousands of civilizations in the universe. Advanced civilizations would look at Earth the same way we look at an ant colony.

Notes

118

Anything that is moving would be warmer than if it were still, and if the universe is expanding, then it would be getting warmer. Yet since we can measure that it's, in fact, getting cooler, then that would mean that the universe isn't moving or expanding.

Notes

119

It's crazy how we eat dead things (plants and animals) to keep us alive. The Source could have created us to eat anything to stay alive (just water, air, sunlight, thoughts, smiles ...anything), yet we eat dead stuff.

It's even crazier to think that we've evolved to the point where we have other people prepare our food for us and even grow it for us, so some humans can spend time doing other things besides survive.

Notes

120

Humans had to have been equipped with certain knowledge. I don't believe that humans evolved to learn how to eat or do anything from the primordial ooze or a paramecium. We innately know how to breathe; no form of life has to try to exist. Some clear parameters have been set within the universe we live in. There are finite rules that this universe adheres to.

Notes

121

The fact that any of this is possible, any physical thing or idea, any concept or action, ...numbers, language, every sensation, EVERYTHING, is completely fascinating to me. I'm in awe that any possible thing even exists. I marvel at it all.

Notes

@ThinKingJuse

Email: ThinKingJUSE@gmail.com

Check out these albums from The JUSE, available on your favorite download and streaming sites:

After Forever

New Religion & New Religion: Instrumentals

Mirror = Glass = Sand = Man
(4 parts)

MK Ultra: Soul or System

Made in the USA
Monee, IL
04 November 2026

67380043R00144